AF588494

JAGUARS

by Elizabeth Andrews

Cody Koala
An Imprint of Pop!
popbooksonline.com

Hello! My name is Cody Koala

This book is filled with videos, puzzles, games, and more! Scan the QR codes* while you read, or visit the website below to make this book pop.

popbooksonline.com/jaguar

*Scanning QR codes requires a web-enabled smart device with a QR code reader app and a camera.

abdobooks.com
Published by Pop!, a division of ABDO, PO Box 398166, Minneapolis, Minnesota 55439.

Printed in the United States of America, North Mankato, Minnesota.
102024
012025
THIS BOOK CONTAINS RECYCLED MATERIALS

Cover Photo: Getty Images
Interior Photos: Getty Images, Shutterstock Images
Editor: Grace Hansen
Series Designer: Neil Klinepier, Candice Keimig

Library of Congress Control Number: 2024938599

Publisher's Cataloging-in-Publication Data
Names: Andrews, Elizabeth, author.
Title: Jaguars / by Elizabeth Andrews
Description: Minneapolis, Minnesota : Pop!, 2025 | Series: Big cats | Includes online resources and index
Identifiers: ISBN 9781098246907 (lib. bdg.) | ISBN 9781098247461 (ebook)
Subjects: LCSH: Big cats--Juvenile literature. | Wildcat--Juvenile literature. | Jaguar--Juvenile literature. | Cats-Behavior--Juvenile literature.
Classification: DDC 599.755--dc23

Table of Contents

Chapter 1

What Is a Jaguar?

The jaguar is the third-largest cat in the world. Jaguars have tannish-orange fur with spots called **rosettes**. Their spots help them **camouflage**. Some jaguars are almost completely black.

Black jaguars are sometimes called black panthers.

Watch a video here!

Jaguars weigh up to 350 lbs (159kg). They are sometimes confused for leopards because of their similar markings. Jaguars are larger, stronger, and have more complex spots than leopards.

Where Do Jaguars Live?

Jaguars live in South America, Central America, and Mexico. Long ago, people **worshipped** jaguars for their strength and fierceness.

Where Jaguars Live

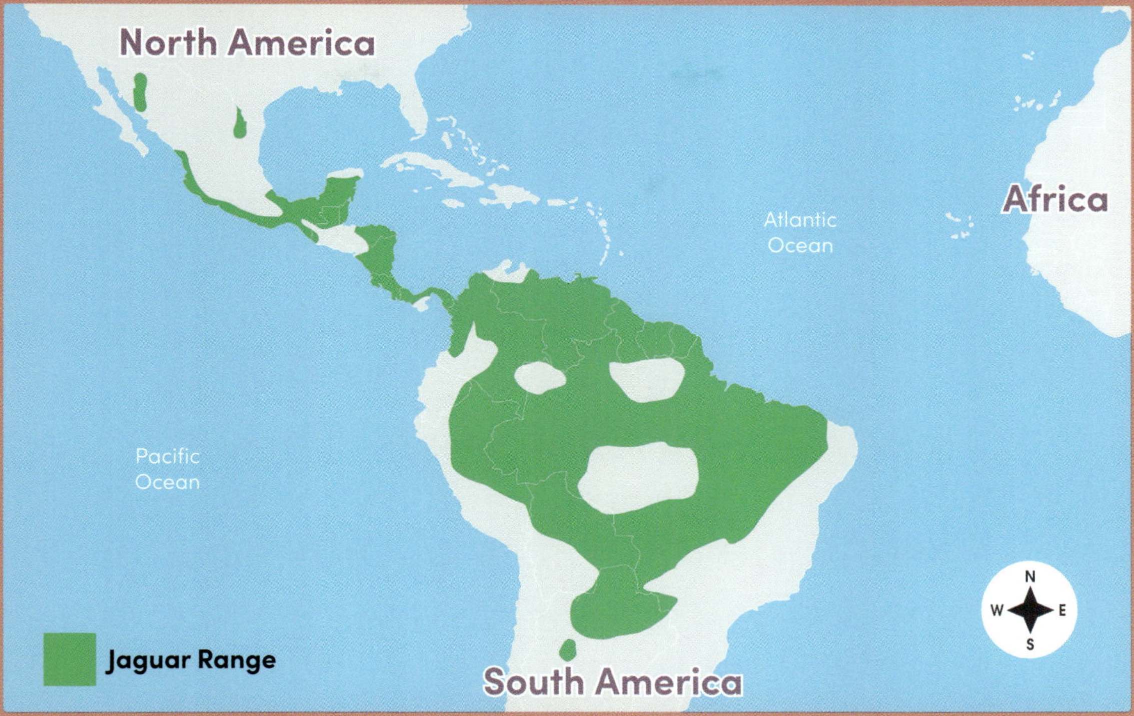

North America
Africa
Atlantic Ocean
Pacific Ocean
Jaguar Range
South America
N
W
E
S

Jaguars usually live in swamps or rainforests. They like to be near lots of plants. They also need to live near water. Unlike most cats, they like to swim.

Jaguars can survive in deserts and grasslands.

Chapter 3

Jaguar Habits

Jaguars are true big cats. They can roar! They also growl, but they cannot purr. Jaguars are **carnivores**. They eat animals such as deer, fish, turtles, and capybaras.

Explore links here!

A jaguar's bite is stronger than a lion's. It can break through a turtle's shell!

Jaguars are excellent swimmers. This helps them hunt for **aquatic** animals. Jaguars usually hunt at night. They closely follow their **prey** and then move quickly to capture it.

Jaguars mostly live alone and have their own **territories**. They use claw and scent marks to let other jaguars know to stay away.

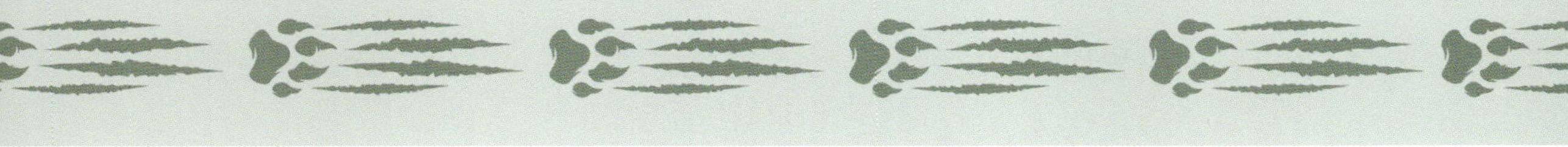

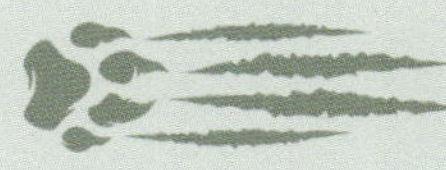

Chapter 4

Jaguar Cubs

Mother jaguars have one to four cubs at a time. They often have twins. Baby jaguars are born with their spots. Their mothers keep them in a den for safety.

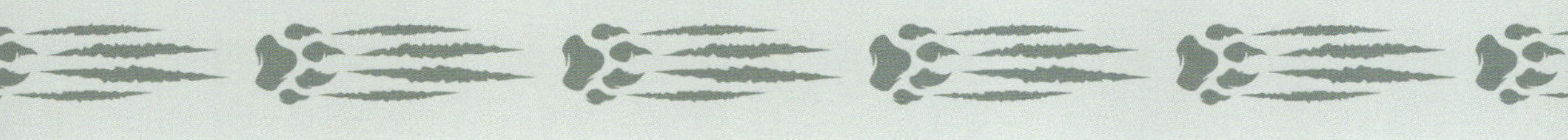

Complete an activity here!

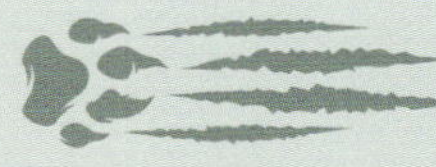

Jaguar cubs start following their mother around when they are six to eight weeks old. They learn from her and stay with her until they are two. Jaguars usually live between 12 and 15 years.

Making Connections

Text-to-Self

What big cat are you most interested in? Please explain your answer.

Text-to-Text

Have you read about any other kinds of big cats? If so, how were those big cats similar to or different from jaguars?

Text-to-World

Ancient people worshipped jaguars. What other animals do you think would have been worshipped in the past? Please explain your answer.

Glossary

aquatic – having to do with water.

camouflage – to hide by coloring or covering to look like the surroundings.

carnivore – an animal that feeds on other animals.

prey – an animal hunted by other animals for food.

rosette – a spot shaped like a rose, often with a dot in the center.

territory – a particular area of land that belongs to an animal.

worship – to show love, respect, and affection to a being.

Index

Online Resources

popbooksonline.com

Thanks for reading this Cody Koala book!

This book is filled with videos, puzzles, games, and more! Scan the QR codes* while you read, or visit the website below to make this book pop.

popbooksonline.com/jaguar

*Scanning QR codes requires a web-enabled smart device with a QR code reader app and a camera.